Chapter 6
Fraction Equivalence and Comparison

Houghton
Mifflin
Harcourt

© Houghton Mifflin Harcourt Publishing Company • Cover Image Credits: (Ground Squirrel) ©Don Johnston/All Canada Photos/Getty Images; (Sawtooth Range, Idaho) ©Ron and Patty Thomas Photography/E+/Getty Images

Made in the United States
Text printed on 100%
recycled paper

Houghton
Mifflin
Harcourt

Copyright © 2015 by Houghton Mifflin Harcourt Publishing Company

All rights reserved. No part of this work may be reproduced or transmitted in any form or by any means, electronic or mechanical, including photocopying or recording, or by any information storage and retrieval system, without the prior written permission of the copyright owner unless such copying is expressly permitted by federal copyright law. Requests for permission to make copies of any part of the work should be addressed to Houghton Mifflin Harcourt Publishing Company, Attn: Contracts, Copyrights, and Licensing, 9400 Southpark Center Loop, Orlando, Florida 32819-8647.

Common Core State Standards © Copyright 2010. National Governors Association Center for Best Practices and Council of Chief State School Offi cers. All rights reserved.

This product is not sponsored or endorsed by the Common Core State Standards Initiative of the National Governors Association Center for Best Practices and the Council of Chief State School Officers.

Printed in the U.S.A.

ISBN 978-0-544-34224-8

27 0877 23

4500864449   C D E F G

If you have received these materials as examination copies free of charge, Houghton Mifflin Harcourt Publishing Company retains title to the materials and they may not be resold. Resale of examination copies is strictly prohibited.

Possession of this publication in print format does not entitle users to convert this publication, or any portion of it, into electronic format.

Dear Students and Families,

Welcome to **Go Math!**, Grade 4! In this exciting mathematics program, there are hands-on activities to do and real-world problems to solve. Best of all, you will write your ideas and answers right in your book. In **Go Math!**, writing and drawing on the pages helps you think deeply about what you are learning, and you will really understand math!

By the way, all of the pages in your **Go Math!** book are made using recycled paper. We wanted you to know that you can Go Green with **Go Math!**

Sincerely,

The Authors

Made in the United States
Text printed on 100% recycled paper

© Houghton Mifflin Harcourt Publishing Company • Image Credits: (bg) ©Sankar Salvady/Flickr/Getty Images; (t) ©Blaine Harrington III/Alamy Images; (c) ©Don Johnston/All Canada Photos/Getty Images; (b) ©Erich Kuchling/Westend61/Corbis

GO MATH!

Authors

Juli K. Dixon, Ph.D.
Professor, Mathematics Education
University of Central Florida
Orlando, Florida

Edward B. Burger, Ph.D.
President, Southwestern University
Georgetown, Texas

Steven J. Leinwand
Principal Research Analyst
American Institutes for
    Research (AIR)
Washington, D.C.

Contributor

Rena Petrello
Professor, Mathematics
Moorpark College
Moorpark, California

Matthew R. Larson, Ph.D.
K-12 Curriculum Specialist for
    Mathematics
Lincoln Public Schools
Lincoln, Nebraska

Martha E. Sandoval-Martinez
Math Instructor
El Camino College
Torrance, California

English Language Learners Consultant

Elizabeth Jiménez
CEO, GEMAS Consulting
Professional Expert on English
    Learner Education
Bilingual Education and
    Dual Language
Pomona, California

© Houghton Mifflin Harcourt Publishing Company • Image Credits: (bg) ©Russ Bishop/Alamy Images ; (t) ©Richard Wear/Design Pics/Corbis

Fractions and Decimals

Critical Area Developing an understanding of fraction equivalence, addition and subtraction of fractions with like denominators, and multiplication of fractions by whole numbers

6 Fraction Equivalence and Comparison **325**

COMMON CORE STATE STANDARDS

4.NF Number and Operations–Fractions
Cluster A Extend understanding of fraction equivalence and ordering.
4.NF.A.1, 4.NF.A.2

Critical Area

Go online! Your math lessons are interactive. Use *i*Tools, Animated Math Models, the Multimedia *e*Glossary, and more.

Chapter 6 Overview

Essential Questions:
- What strategies can you use to compare fractions and write equivalent fractions?
- What models can help you compare and order fractions?
- How can you find equivalent fractions?
- How can you solve problems that involve fractions?

Personal Math Trainer
Online Assessment and Intervention

© Houghton Mifflin Harcourt Publishing Company

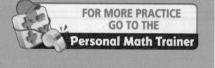

FOR MORE PRACTICE
GO TO THE
Personal Math Trainer

Practice and Homework

Lesson Check and
Spiral Review in
every lesson

© Houghton Mifflin Harcourt Publishing Company

Critical Area

Fractions and Decimals

Common Core

CRITICAL AREA Developing an understanding of fraction equivalence, addition and subtraction of fractions with like denominators, and multiplication of fractions by whole numbers

A *luthier,* or guitar maker, at his workshop

© Houghton Mifflin Harcourt Publishing Company • Image Credits: (inset) ©Danny Lehman/Corbis

Building Custom Guitars

Do you play the guitar, or would you like to learn how to play one? The guitar size you need depends on your height to the nearest inch and on *scale length*. Scale length is the distance from the *bridge* of the guitar to the *nut*.

Get Started   WRITE ▸ Math

Order the guitar sizes from the least size to the greatest size, and complete the table.

Important Facts

Guitar Sizes for Students			
Age of Player	Height of Player (to nearest inch)	Scale Length (shortest to longest, in inches)	Size of Guitar
4–6	3 feet 3 inches to 3 feet 9 inches	19	
6–8	3 feet 10 inches to 4 feet 5 inches	20.5	
8–11	4 feet 6 inches to 4 feet 11 inches	22.75	
11–Adult	5 feet or taller	25.5	

Size of Guitar: $\frac{1}{2}$ size, $\frac{4}{4}$ size, $\frac{1}{4}$ size, $\frac{3}{4}$ size

Adults play $\frac{4}{4}$-size guitars. You can see that guitars also come in $\frac{3}{4}$, $\frac{1}{2}$, and $\frac{1}{4}$ sizes. Figure out which size guitar you would need according to your height and the scale length for each size guitar. Use the Important Facts to decide. **Explain** your thinking.

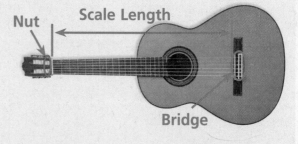

_____

_____

Completed by _____

© Houghton Mifflin Harcourt Publishing Company • Image Credits: ©PhotoDisc/Getty Images

Fraction Equivalence and Comparison

© Houghton Mifflin Harcourt Publishing Company • Image Credits: (b) ©Stocktrek/Brand X Pictures/Getty Images

✓ Show What You Know

Personal Math Trainer
Online Assessment and Intervention

Check your understanding of important skills.

Name _____

▶ **Part of a Whole** Write a fraction for the shaded part. (3.NF.A.1)

1. _____

2. _____

3. _____

▶ **Name the Shaded Part** Write a fraction for the shaded part. (3.NF.A.1)

4. _____

5. _____

6. _____

▶ **Compare Parts of a Whole** Color the fraction strips to show the fractions. Circle the greater fraction. (3.NF.A.3d)

7. $\frac{1}{2}$

$\frac{1}{3}$

8. $\frac{1}{5}$

$\frac{1}{3}$

Math in the Real World

Earth's surface is covered by more than 57 million square miles of land. The table shows about how much of Earth's land surface each continent covers. Which continent covers the greatest part of Earth's land surface?

Continent	Part of Land Surface
Asia	$\frac{3}{10}$
Africa	$\frac{1}{5}$
Antarctica	$\frac{9}{100}$
Australia	$\frac{6}{100}$
Europe	$\frac{7}{100}$
North America	$\frac{1}{6}$
South America	$\frac{1}{8}$

▶ **Visualize It** •

Complete the flow map by using the words with a ✓.

Whole Numbers and Fractions

What is it?   What are some examples?

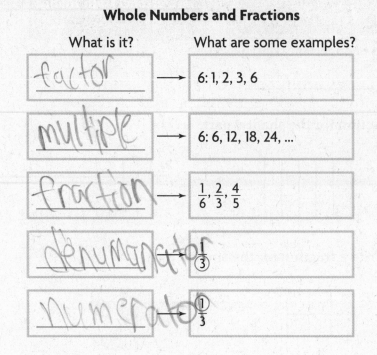

factor ⟶ 6: 1, 2, 3, 6

multiple ⟶ 6: 6, 12, 18, 24, ...

fraction ⟶ $\frac{1}{6}, \frac{2}{3}, \frac{4}{5}$

denumonator ⟶ $\frac{1}{③}$

numerator ⟶ $\frac{①}{3}$

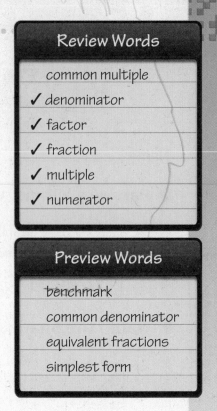

Review Words
common multiple
✓ denominator
✓ factor
✓ fraction
✓ multiple
✓ numerator

Preview Words
benchmark
common denominator
equivalent fractions
simplest form

▶ **Understand Vocabulary** •

Complete the sentences by using preview words.

1. A fraction is in _____ if the numerator and denominator have only 1 as a common factor.

2. _____ name the same amount.

3. A _____ is a common multiple of two or more denominators.

4. A _____ is a known size or amount that helps you understand a different size or amount.

GO DIGITAL
• Interactive Student Edition
• Multimedia eGlossary

© Houghton Mifflin Harcourt Publishing Company

Chapter 6 Vocabulary

benchmark

punto de referencia

6

common denominator

denominador común

9

denominator

denominador

22

equivalent fractions

fracciones equivalentes

29

fraction

fracción

36

multiple

múltiplo

55

numerator

numerador

56

simplest form

mínima expresión

84

© Houghton Mifflin Harcourt Publishing Company

A common multiple of two or more denominators

Example: Some common denominators for $\frac{1}{4}$ and $\frac{5}{6}$ are 12, 24, and 36.

© Houghton Mifflin Harcourt Publishing Company

A known size or amount that helps you understand a different size or amount

You can use $\frac{1}{2}$ as a benchmark to help you compare fractions.

© Houghton Mifflin Harcourt Publishing Company

Two or more fractions that name the same amount

Example: $\frac{3}{4}$ and $\frac{6}{8}$ name the same amount.

$$\frac{3}{4} = \frac{6}{8}$$

© Houghton Mifflin Harcourt Publishing Company

The number below the bar in a fraction that tells how many equal parts are in the whole or in the group

Example: $\frac{3}{4}$ ⟵ denominator

© Houghton Mifflin Harcourt Publishing Company

The product of a number and a counting number is called a multiple of the number

Example:

$$\begin{array}{cccc} 3 & 3 & 3 & 3 \\ \times 1 & \times 2 & \times 3 & \times 4 \\ \hline 3 & 6 & 9 & 12 \end{array}$$

⟵ counting numbers
⟵ multiples of 3

© Houghton Mifflin Harcourt Publishing Company

A number that names a part of a whole or part of a group

Example:

$\frac{1}{3}$

© Houghton Mifflin Harcourt Publishing Company

A fraction is in simplest form when the numerator and denominator have only 1 as a common factor

$\frac{2}{8} = \frac{1}{4}$

simplest form

© Houghton Mifflin Harcourt Publishing Company

The number above the bar in a fraction that tells how many parts of the whole or group are being considered

Example: $\frac{1}{5}$ ⟵ numerator

Going to San Francisco

For 2 to 4 players

Word Box
- benchmark
- common denominator
- denominator
- equivalent fractions
- fraction
- multiple
- numerator
- simplest form

Materials
- 3 of one color per player: red, blue, green, and yellow playing pieces
- 1 number cube

How to Play

1. Put your 3 playing pieces in the START circle of the same color.

2. To get a playing piece out of START, you must toss a 6.
   - If you toss a 6, move 1 of your playing pieces to the same-colored circle on the path.
   - If you do not toss a 6, wait until your next turn.

3. Once you have a playing piece on the path, toss the number cube to take a turn. Move the playing piece that many tan spaces. You must get all three of your playing pieces on the path.

4. If you land on a space with a question, answer it. If you are correct, move ahead 1 space.

5. To reach FINISH, you must move your playing piece up the path that is the same color as your playing piece. The first player to get all three playing pieces on FINISH wins.

CALIFORNIA

© Houghton Mifflin Harcourt Publishing Company • ©Brand X Pictures/Getty Images; (inset) ©Liquid Library/Jupiterimages/Getty Images

START

START

Explain how you know that $\frac{2}{12}$ is in simplest form.

How do you know when a fraction is in simplest form?

What is a fraction?

What are equivalent fractions?

FINISH

How can you find a fraction that is equivalent to $\frac{1}{3}$?

Why are $\frac{1}{2}$ and $\frac{2}{4}$ equivalent fractions?

What is the difference between a multiple and common multiple?

Explain the relationship between a common multiple and a common denominator.

© Houghton Mifflin Harcourt Publishing Company • (bg) ©Digital Vision/Getty Images; (tl) ©Getty Images; (bl)

Game

START

What is the meaning of common denominator?

What is the simplest form of a fraction?

How do you find the common denominator of $\frac{2}{3}$ and $\frac{3}{5}$?

What is a multiple?

FINISH

What is a benchmark?

How are benchmarks used to compare fractions?

In a fraction, what does the numerator represent?

In a fraction, what does the denominator represent?

START

FISHERMANS WHARF · OF SAN FRANC

© Houghton Mifflin Harcourt Publishing Company • (tr) ©PhotoDisc/Getty Images

The Write Way

Reflect

Explain how to find the common denominator for $\frac{2}{3}$ and $\frac{1}{4}$.

• Work with a partner to explain and illustrate two ways to find equivalent fractions. Use a separate piece of paper for your drawing.

• Write about all the different ways you can show 25.

• Summarize how you would order the fractions $\frac{2}{3}$, $\frac{7}{8}$, and $\frac{4}{5}$, including any "false starts" or "dead ends."

© Houghton Mifflin Harcourt Publishing Company • ©Digital Vision/Getty Images

Name _____

Equivalent Fractions

Essential Question How can you use models to show equivalent fractions?

Common Core

Number and Operations— Fractions—4.NF.A.1

MATHEMATICAL PRACTICES
MP4, MP5, MP7

Hands On

Investigate

Materials ■ color pencils

Joe cut a pan of lasagna into third-size pieces. He kept $\frac{1}{3}$ and gave the rest away. Joe will not eat his part all at once. How can he cut his part into smaller, equal-size pieces?

A. Draw on the model to show how Joe could cut his part of the lasagna into 2 equal pieces.

You can rename these 2 equal pieces as a fraction of the original pan of lasagna.

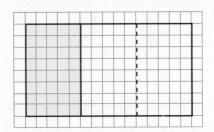

Suppose Joe had cut the original pan of lasagna into equal pieces of this size.

How many pieces would there be? _____

What fraction of the pan is 1 piece? _____

What fraction of the pan is 2 pieces? _____

You can rename $\frac{1}{3}$ as _____.

B. Now draw on the model to show how Joe could cut his part of the lasagna into 4 equal pieces.

You can rename these 4 equal pieces as a fraction of the original pan of lasagna.

Suppose Joe had cut the original pan of lasagna into equal pieces of this size.

How many pieces would there be? _____

What fraction of the pan is 1 piece? _____

What fraction of the pan is 4 pieces? _____

You can rename $\frac{1}{3}$ as _____.

C. Fractions that name the same amount are **equivalent fractions**. Write the equivalent fractions.

$$\frac{1}{3} = \underline{} = \underline{}$$

© Houghton Mifflin Harcourt Publishing Company

Draw Conclusions

1. Compare the models for $\frac{1}{3}$ and $\frac{2}{6}$. How does the number of parts relate to the sizes of the parts?

_____

_____

2. Describe how the numerators are related and how the denominators are related in $\frac{1}{3} = \frac{2}{6}$.

_____

_____

_____

3. **THINK SMARTER** Does $\frac{1}{3} = \frac{3}{9}$? Explain.

_____

_____

Make Connections

Savannah has $\frac{2}{4}$ yard of ribbon, and Isabel has $\frac{3}{8}$ yard of ribbon. How can you determine whether Savannah and Isabel have the same length of ribbon?

The equal sign ($=$) and not equal to sign ($\neq$) show whether fractions are equivalent.

Tell whether $\frac{2}{4}$ and $\frac{3}{8}$ are equivalent. Write $=$ or $\neq$.

STEP 1 Shade the amount of ribbon Savannah has.

$$\frac{0}{4} \quad \frac{1}{4} \quad \frac{2}{4} \quad \frac{3}{4} \quad \frac{4}{4}$$

STEP 2 Shade the amount of ribbon Isabel has.

$$\frac{0}{8} \quad \frac{1}{8} \quad \frac{2}{8} \quad \frac{3}{8} \quad \frac{4}{8} \quad \frac{5}{8} \quad \frac{6}{8} \quad \frac{7}{8} \quad \frac{8}{8}$$

Think: $\frac{2}{4}$ yard is not the same amount as $\frac{3}{8}$ yard.

So, $\frac{2}{4} \bigcirc \frac{3}{8}$.

Math Talk

MATHEMATICAL PRACTICES ④

Use Models How could you use a model to show that $\frac{4}{8} = \frac{1}{2}$?

© Houghton Mifflin Harcourt Publishing Company

Name _____

Use the model to write an equivalent fraction.

1.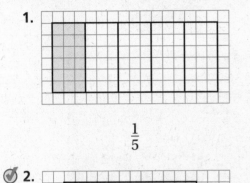

$$\frac{1}{5} \qquad = \qquad \underline{\hspace{2cm}}$$

✓ 2.

$$\frac{2}{3} \qquad = \qquad \underline{\hspace{2cm}}$$

Tell whether the fractions are equivalent. Write = or ≠.

3. $\frac{1}{6} \bigcirc \frac{2}{12}$ 4. $\frac{2}{5} \bigcirc \frac{6}{10}$ 5. $\frac{4}{12} \bigcirc \frac{1}{3}$

✓ 6. $\frac{5}{8} \bigcirc \frac{2}{4}$ 7. $\frac{5}{6} \bigcirc \frac{10}{12}$ 8. $\frac{1}{2} \bigcirc \frac{5}{10}$

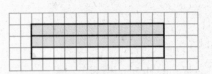

Problem Solving • Applications

9. **GO DEEPER** Manny used 8 tenth-size parts to model $\frac{8}{10}$. Ana used fewer parts to model an equivalent fraction. How does the size of a part in Ana's model compare to the size of a tenth-size part? What size part did Ana use?

_____

_____

_____

10. **MATHEMATICAL PRACTICE ⑤ Use a Concrete Model** How many eighth-size parts do you need to model $\frac{3}{4}$? Explain.

_____

_____

_____

_____

© Houghton Mifflin Harcourt Publishing Company

What's the Error?

11. **THINK SMARTER** Ben brought two pizzas to a party. He says that since $\frac{1}{4}$ of each pizza is left, the same amount of each pizza is left. What is his error?

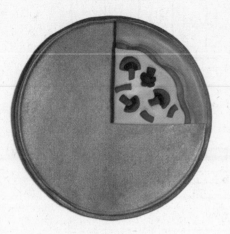

Draw models of 2 pizzas with a different number of equal pieces. Use shading to show $\frac{1}{4}$ of each pizza.

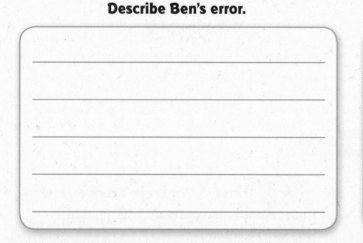

Describe Ben's error.

12. **THINK SMARTER** For numbers 12a–12d, tell whether the fractions are equivalent by selecting the correct symbol.

12a. $\dfrac{3}{15}$ $\begin{array}{c} = \\ \neq \end{array}$ $\dfrac{1}{6}$

12b. $\dfrac{3}{4}$ $\begin{array}{c} = \\ \neq \end{array}$ $\dfrac{16}{20}$

12c. $\dfrac{2}{3}$ $\begin{array}{c} = \\ \neq \end{array}$ $\dfrac{8}{12}$

12d. $\dfrac{8}{10}$ $\begin{array}{c} = \\ \neq \end{array}$ $\dfrac{4}{5}$

© Houghton Mifflin Harcourt Publishing Company

Equivalent Fractions

 COMMON CORE STANDARD—4.NF.A.1
Extend understanding of fraction equivalence and ordering.

Use the model to write an equivalent fraction.

1.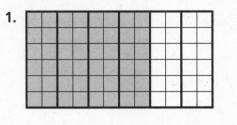

$\frac{4}{6}$ =   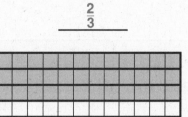 $\frac{2}{3}$

2.

$\frac{3}{4}$ =   _____

Tell whether the fractions are equivalent. Write = or ≠.

3. $\frac{8}{10} \bigcirc \frac{4}{5}$ **4.** $\frac{1}{2} \bigcirc \frac{7}{12}$ **5.** $\frac{3}{4} \bigcirc \frac{8}{12}$ **6.** $\frac{2}{3} \bigcirc \frac{4}{6}$

Problem Solving

7. Jamal finished $\frac{5}{6}$ of his homework. Margaret finished $\frac{3}{4}$ of her homework, and Steve finished $\frac{10}{12}$ of his homework. Which two students finished the same amount of homework?

8. Sophia's vegetable garden is divided into 12 equal sections. She plants carrots in 8 of the sections. Write two fractions that are equivalent to the part of Sophia's garden that is planted with carrots.

_____   _____

9. **WRITE** ▸*Math* Draw a model to show a fraction that is equivalent to $\frac{1}{3}$ and a fraction that is not equivalent to $\frac{1}{3}$.

© Houghton Mifflin Harcourt Publishing Company

Lesson Check (4.NF.A.1)

1. A rectangle is divided into 8 equal parts. Two parts are shaded. What fraction is equivalent to the shaded area of the rectangle?

2. Jeff uses 3 fifth-size strips to model $\frac{3}{5}$. He wants to use tenth-size strips to model an equivalent fraction. How many tenth-size strips will he need?

_____

_____

Spiral Review (4.OA.A.3, 4.OA.B.4, 4.NBT.B.5, 4.NBT.B.6)

3. Cassidy places 40 stamps on each of 8 album pages. How many stamps does she place?

4. Maria and 3 friends have 1,200 soccer cards. If they share the soccer cards equally, how many will each person receive?

_____

_____

5. Six groups of students sell 162 balloons at the school carnival. There are 3 students in each group. If each student sells the same number of balloons, how many balloons does each student sell?

6. Four students each made a list of prime numbers.

   Eric: 5, 7, 17, 23
   Maya: 3, 5, 13, 17
   Bella: 2, 3, 17, 19
   Jordan: 7, 11, 13, 21

Who made an error and included a composite number? Write the composite number from his or her list.

_____

_____

© Houghton Mifflin Harcourt Publishing Company

FOR MORE PRACTICE
GO TO THE
Personal Math Trainer

Name _____

Generate Equivalent Fractions

Essential Question How can you use multiplication to find equivalent fractions?

Common Core **Number and Operations—Fractions—4.NF.A.1**

MATHEMATICAL PRACTICES
MP4, MP7, MP8

Unlock the Problem (Real World)

Sara needs $\frac{3}{4}$ cup of dish soap to make homemade bubble solution. Her measuring cup is divided into eighths. What fraction of the measuring cup should Sara fill with dish soap?

- Is an eighth-size part of a measuring cup bigger or smaller than a fourth-size part?

_____

Find how many eighths are in $\frac{3}{4}$.

STEP 1 Compare fourths and eighths.

Shade to model $\frac{1}{4}$.
Use fourth-size parts.

1 part

Shade to model $\frac{1}{4}$.
Use eighth-size parts.

2 parts

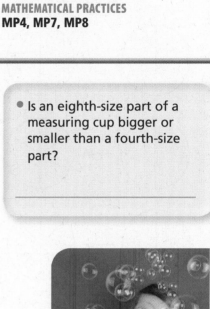

You need _____ eighth-size parts to make 1 fourth-size part.

STEP 2 Find how many eighths you need to make 3 fourths.

Shade to model $\frac{3}{4}$.
Use fourth-size parts.

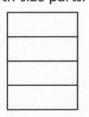
3 parts

Shade to model $\frac{3}{4}$.
Use eighth-size parts.

6 parts

You needed 2 eighth-size parts to make 1 fourth-size part.

So, you need _____ eighth-size parts to make 3 fourth-size parts.

So, Sara should fill $\frac{}{8}$ of the measuring cup with dish soap.

Math Talk

MATHEMATICAL PRACTICES ④

Interpret a Result Explain how you knew the number of eighth-size parts you needed to make 1 fourth-size part.

1. Explain why 6 eighth-size parts is the same amount as 3 fourth-size parts.

_____

© Houghton Mifflin Harcourt Publishing Company • Image Credits: (t) ©Muriel de Seze/Getty Images

🔓 Example Write four fractions that are equivalent to $\frac{1}{2}$.

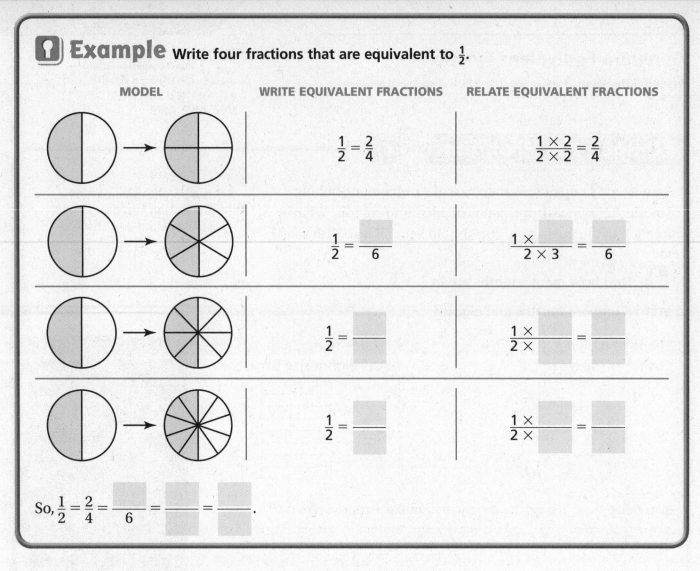

MODEL	WRITE EQUIVALENT FRACTIONS	RELATE EQUIVALENT FRACTIONS
	$\frac{1}{2} = \frac{2}{4}$	$\frac{1 \times 2}{2 \times 2} = \frac{2}{4}$
	$\frac{1}{2} = \frac{}{6}$	$\frac{1 \times }{2 \times 3} = \frac{}{6}$
	$\frac{1}{2} = $	$\frac{1 \times }{2 \times } = $
	$\frac{1}{2} = $	$\frac{1 \times }{2 \times } = $

So, $\frac{1}{2} = \frac{2}{4} = \frac{}{6} = \underline{} = \underline{}$.

2. Look at the model that shows $\frac{1}{2} = \frac{3}{6}$. How does the number of parts in the whole affect the number of parts that are shaded? Explain.

_____

_____

3. Explain how you can use multiplication to write a fraction that is equivalent to $\frac{3}{5}$.

_____

_____

_____

4. Are $\frac{2}{3}$ and $\frac{6}{8}$ equivalent? Explain.

_____

© Houghton Mifflin Harcourt Publishing Company

Name _____

1. Complete the table below.

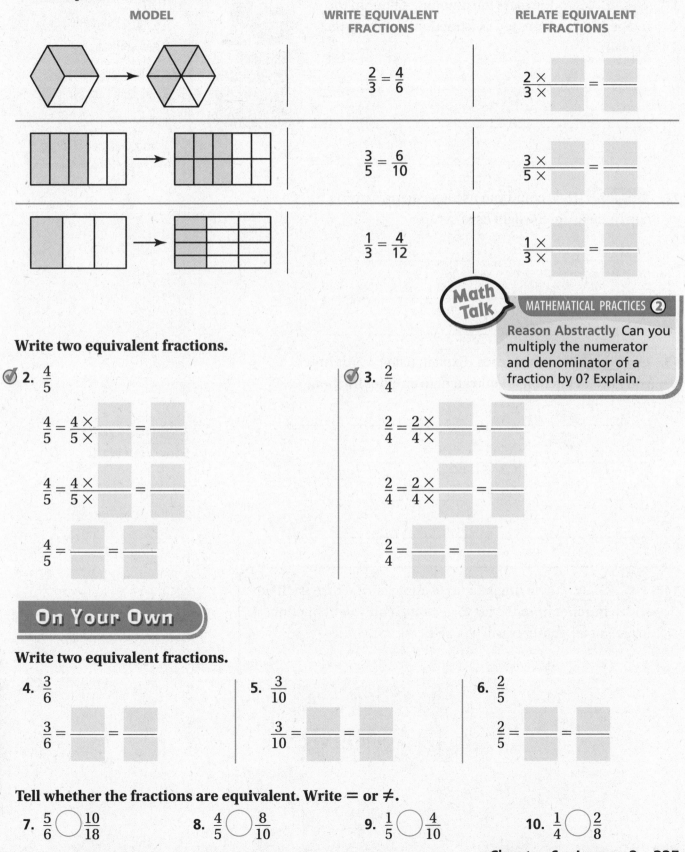

MODEL	WRITE EQUIVALENT FRACTIONS	RELATE EQUIVALENT FRACTIONS
	$\frac{2}{3} = \frac{4}{6}$	$\frac{2 \times }{3 \times } = \frac{}{}$
	$\frac{3}{5} = \frac{6}{10}$	$\frac{3 \times }{5 \times } = \frac{}{}$
	$\frac{1}{3} = \frac{4}{12}$	$\frac{1 \times }{3 \times } = \frac{}{}$

Math Talk · MATHEMATICAL PRACTICES ②

Reason Abstractly Can you multiply the numerator and denominator of a fraction by 0? Explain.

Write two equivalent fractions.

☑ **2.** $\frac{4}{5}$

$\frac{4}{5} = \frac{4 \times }{5 \times } = \frac{}{}$

$\frac{4}{5} = \frac{4 \times }{5 \times } = \frac{}{}$

$\frac{4}{5} = \frac{}{} = \frac{}{}$

☑ **3.** $\frac{2}{4}$

$\frac{2}{4} = \frac{2 \times }{4 \times } = \frac{}{}$

$\frac{2}{4} = \frac{2 \times }{4 \times } = \frac{}{}$

$\frac{2}{4} = \frac{}{} = \frac{}{}$

Write two equivalent fractions.

4. $\frac{3}{6}$

$\frac{3}{6} = \frac{}{} = \frac{}{}$

5. $\frac{3}{10}$

$\frac{3}{10} = \frac{}{} = \frac{}{}$

6. $\frac{2}{5}$

$\frac{2}{5} = \frac{}{} = \frac{}{}$

Tell whether the fractions are equivalent. Write = or ≠.

7. $\frac{5}{6} \bigcirc \frac{10}{18}$

8. $\frac{4}{5} \bigcirc \frac{8}{10}$

9. $\frac{1}{5} \bigcirc \frac{4}{10}$

10. $\frac{1}{4} \bigcirc \frac{2}{8}$

© Houghton Mifflin Harcourt Publishing Company

Problem Solving • Applications Real World

Use the recipe for 11–12.

11. THINK SMARTER Kim says the amount of flour in the recipe can be expressed as a fraction. Is she correct? Explain.

_____

_____

Face Paint Recipe

$\frac{2}{8}$ cup cornstarch

1 tablespoon flour

$\frac{9}{12}$ cup light corn syrup

$\frac{1}{4}$ cup water

$\frac{1}{2}$ teaspoon food coloring

12. GO DEEPER How could you use a $\frac{1}{8}$-cup measuring cup to measure the light corn syrup?

_____

_____

13. MATHEMATICAL PRACTICE ⑤ **Communicate** Explain using words how you know a fraction is equivalent to another fraction.

_____

_____

_____

WRITE Math
Show Your Work

14. THINK SMARTER Kyle drank $\frac{2}{3}$ cup of apple juice. Fill in each box with a number from the list to generate equivalent fractions for $\frac{2}{3}$. Not all numbers will be used.

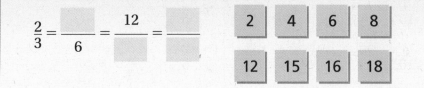

$$\frac{2}{3} = \frac{}{6} = \frac{12}{} = \frac{}{}$$

| 2 | 4 | 6 | 8 |
| 12 | 15 | 16 | 18 |

© Houghton Mifflin Harcourt Publishing Company

Generate Equivalent Fractions

Common Core

COMMON CORE STANDARD—4.NF.A.1
Extend understanding of fraction equivalence and ordering.

Write two equivalent fractions for each.

1. $\frac{1}{3}$

$$\frac{1 \times 2}{3 \times 2} = \frac{2}{6}$$

$$\frac{1 \times 4}{3 \times 4} = \frac{4}{12}$$

2. $\frac{2}{3}$

_____

3. $\frac{1}{2}$

_____

4. $\frac{4}{5}$

_____

Tell whether the fractions are equivalent.
Write = or ≠.

5. $\frac{1}{4} \bigcirc \frac{3}{12}$

6. $\frac{4}{5} \bigcirc \frac{5}{10}$

7. $\frac{3}{8} \bigcirc \frac{2}{6}$

8. $\frac{3}{4} \bigcirc \frac{6}{8}$

9. $\frac{5}{6} \bigcirc \frac{10}{12}$

10. $\frac{6}{12} \bigcirc \frac{5}{8}$

11. $\frac{2}{5} \bigcirc \frac{4}{10}$

12. $\frac{2}{4} \bigcirc \frac{3}{12}$

Problem Solving · Real World

13. Jan has a 12-ounce milkshake. Four ounces in the milkshake are vanilla, and the rest is chocolate. What are two equivalent fractions that represent the fraction of the milkshake that is vanilla?

14. Kareem lives $\frac{4}{10}$ of a mile from the mall. Write two equivalent fractions that show what fraction of a mile Kareem lives from the mall.

_____

_____

15. **WRITE** ▸*Math* Explain how you can determine if $\frac{1}{3}$ and $\frac{4}{12}$ are equivalent fractions.

_____

_____

© Houghton Mifflin Harcourt Publishing Company

Lesson Check (4.NF.A.1)

1. Jessie colored a poster. She colored $\frac{2}{5}$ of the poster red. Write a fraction that is equivalent to $\frac{2}{5}$.

_____

2. Marcus makes a punch that is $\frac{1}{4}$ cranberry juice. Write two fractions that are equivalent to $\frac{1}{4}$.

_____

Spiral Review (4.OA.A.3, 4.OA.C.5, 4.NBT.B.5)

3. An electronics store sells a large flat screen television for $1,699. Last month, the store sold 8 of these television sets. About how much money did the televisions sell for?

_____

4. Matthew has 18 sets of baseball cards. Each set has 12 cards. About how many baseball cards does Matthew have?

_____

5. Diana had 41 stickers. She put them in 7 equal groups. She put as many as possible in each group. She gave the leftover stickers to her sister. How many stickers did Diana give to her sister?

_____

6. Christopher wrote the number pattern below. The first term is 8.

8, 6, 9, 7, 10, ...

What is a rule for the pattern?

_____

© Houghton Mifflin Harcourt Publishing Company

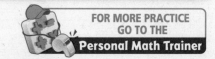

FOR MORE PRACTICE
GO TO THE
Personal Math Trainer

Name _____

Simplest Form

Essential Question How can you write a fraction as an equivalent fraction in simplest form?

Common Core **Number and Operations— Fractions—4.NF.A.1**
MATHEMATICAL PRACTICES
MP4, MP6, MP7

⚷ Unlock the Problem Real World

Vicki made a fruit tart and cut it into 6 equal pieces. Vicki, Silvia, and Elena each took 2 pieces of the tart home. Vicki says she and each of her friends took $\frac{1}{3}$ of the tart home. Is Vicki correct?

- Into how many pieces was the tart cut?

_____

- How many pieces did each girl take?

_____

🔑 Activity

Materials ■ color pencils

STEP 1 Use a blue pencil to shade the pieces Vicki took home.

_____

STEP 2 Use a red pencil to shade the pieces Silvia took home.

_____

STEP 3 Use a yellow pencil to shade the pieces Elena took home.

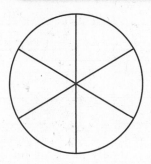

The tart is divided into _____ equal-size pieces. The 3 colors on the model show how to combine sixth-size pieces to make

_____ equal third-size pieces.

So, Vicki is correct. Vicki, Silvia, and Elena each took —— of the tart home.

Math Talk **MATHEMATICAL PRACTICES ④**

Interpret a Result
Compare the models for $\frac{2}{6}$ and $\frac{1}{3}$. Explain how the sizes of the parts are related.

- What if Vicki took 3 pieces of the tart home and Elena took 3 pieces of the tart home. How could you combine the pieces to write a fraction that represents the part each friend took home? Explain.

_____

_____

_____

© Houghton Mifflin Harcourt Publishing Company

Simplest Form A fraction is in **simplest form** when you can represent it using as few equal parts of a whole as possible. You need to describe the part you have in equal-size parts. If you can't describe the part you have using fewer parts, then you cannot simplify the fraction.

One Way Use models to write an equivalent fraction in simplest form.

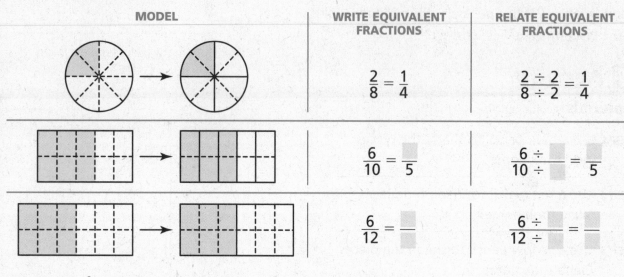

MODEL	WRITE EQUIVALENT FRACTIONS	RELATE EQUIVALENT FRACTIONS
	$\dfrac{2}{8} = \dfrac{1}{4}$	$\dfrac{2 \div 2}{8 \div 2} = \dfrac{1}{4}$
	$\dfrac{6}{10} = \dfrac{\square}{5}$	$\dfrac{6 \div \square}{10 \div \square} = \dfrac{\square}{5}$
	$\dfrac{6}{12} = \dfrac{\square}{\square}$	$\dfrac{6 \div \square}{12 \div \square} = \dfrac{\square}{\square}$

To simplify $\frac{6}{10}$, you can combine tenth-size parts into equal groups with 2 parts each.

So, $\dfrac{6}{10} = \dfrac{6 \div \square}{10 \div \square} = \dfrac{\square}{\square}$.

Another Way Use common factors to write $\frac{6}{10}$ in simplest form.

A fraction is in simplest form when 1 is the only factor that the numerator and denominator have in common. The parts of the whole cannot be combined into fewer equal-size parts to show the same fraction.

STEP 1 List the factors of the numerator and denominator. Circle common factors.

Factors of 6: _____, _____, _____, _____

Factors of 10: _____, _____, _____, _____

STEP 2 Divide the numerator and denominator by a common factor greater than 1.

$\dfrac{6}{10} = \dfrac{6 \div \square}{10 \div \square} = \dfrac{\square}{\square}$

Since 1 is the only factor that 3 and 5 have in common, _____ is written in simplest form.

© Houghton Mifflin Harcourt Publishing Company

Name _____

Simplest Form

COMMON CORE STANDARD—4.NF.A.1
Extend understanding of fraction equivalence and ordering.

Write the fraction in simplest form.

1. $\frac{6}{10}$ 　　2. $\frac{6}{8}$ 　　3. $\frac{5}{5}$ 　　4. $\frac{8}{12}$

$\frac{6}{10} = \frac{6 \div 2}{10 \div 2} = \frac{3}{5}$ _____   _____   _____

5. $\frac{100}{100}$ 　　6. $\frac{2}{6}$ 　　7. $\frac{2}{8}$ 　　8. $\frac{4}{10}$

_____   _____   _____   _____

Tell whether the fractions are equivalent.
Write = or ≠.

9. $\frac{6}{12} \bigcirc \frac{1}{12}$ 10. $\frac{3}{4} \bigcirc \frac{5}{6}$ 11. $\frac{6}{10} \bigcirc \frac{3}{5}$ 12. $\frac{3}{12} \bigcirc \frac{1}{3}$

Problem Solving · Real World

13. At Memorial Hospital, 9 of the 12 babies born on Tuesday were boys. In simplest form, what fraction of the babies born on Tuesday were boys?

14. Cristina uses a ruler to measure the length of her math textbook. She says that the book is $\frac{4}{10}$ meter long. Is her measurement in simplest form? If not, what is the length of the book in simplest form?

_____   _____

15. **WRITE** ▸ Math Explain using words or drawings how to write $\frac{6}{9}$ in simplest form.

_____

_____

_____

© Houghton Mifflin Harcourt Publishing Company

Lesson Check (4.NF.A.1)

1. Six out of the 12 members of the school choir are boys. In simplest form, what fraction of the choir is boys?

2. Write $\frac{10}{12}$ in simplest form.

_____

_____

Spiral Review (4.OA.A.3, 4.OA.B.4, 4.NBT.B.5, 4.NF.A.1)

3. Each of the 23 students in Ms. Evans' class raised $45 for the school by selling coupon books. How much money did the class raise?

4. List two common factors of 36 and 48.

_____

_____

_____

_____

5. Bart uses $\frac{3}{12}$ cup milk to make muffins. Write a fraction that is equivalent to $\frac{3}{12}$.

6. Ashley bought 4 packages of juice boxes. There are 6 juice boxes in each package. She gave 2 juice boxes to each of 3 friends. How many juice boxes does Ashley have left?

_____

_____

© Houghton Mifflin Harcourt Publishing Company

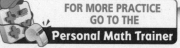

FOR MORE PRACTICE
GO TO THE
Personal Math Trainer

Name _____

Common Denominators

Essential Question How can you write a pair of fractions as fractions with a common denominator?

Common Core Number and Operations—Fractions—4.NF.A.1
MATHEMATICAL PRACTICES
MP4, MP6, MP7

Unlock the Problem Real World

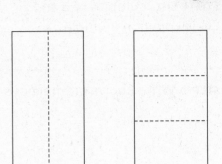

Martin has two rectangles that are the same size. One rectangle is cut into $\frac{1}{2}$-size parts. The other rectangle is cut into $\frac{1}{3}$-size parts. He wants to cut the rectangles so they have the same size parts. How can he cut each rectangle?

A **common denominator** is a common multiple of the denominators of two or more fractions. Fractions with common denominators represent wholes cut into the same number of parts.

Activity Use paper folding and shading.

Materials ■ 2 sheets of paper

Find a common denominator for $\frac{1}{2}$ and $\frac{1}{3}$.

STEP 1

Model the rectangle cut into $\frac{1}{2}$-size parts. Fold one sheet of paper in half. Draw a line on the fold.

_____

STEP 2

Model the rectangle cut into $\frac{1}{3}$-size parts. Fold the other sheet of paper into thirds. Draw lines on the folds.

_____

STEP 3

Fold each sheet of paper so that both sheets have the same number of parts. Draw lines on the folds. How many equal

parts does each sheet of paper have? _____

STEP 4

Draw a picture of your sheets of paper to show how many parts each rectangle could have.

So, each rectangle could be cut into _____ parts.

MATHEMATICAL PRACTICES ④

Use Models How did the models help you find the common denominator for $\frac{1}{2}$ and $\frac{1}{3}$?

© Houghton Mifflin Harcourt Publishing Company

🔒 Example Write $\frac{4}{5}$ and $\frac{1}{2}$ as a pair of fractions with common denominators.

You can use common multiples to find a common denominator. List multiples of each denominator. A common multiple can be used as a common denominator.

STEP 1 List multiples of 5 and 2. Circle common multiples.

5:    5,   10,   _____, _____, _____, _____

2:    _____, _____, _____, _____, _____, _____

STEP 2 Write equivalent fractions.

$$\frac{4}{5} = \frac{4 \times }{5 \times } = \frac{}{10}$$

$$\frac{1}{2} = \frac{1 \times }{2 \times } = \frac{}{10}$$

Choose a denominator that is a common multiple of 5 and 2.

You can write $\frac{4}{5}$ and $\frac{1}{2}$ as _____ and _____ .

> **⚠ ERROR Alert**
>
> Remember that when you multiply the denominator by a factor, you must multiply the numerator by the same factor to write an equivalent fraction.

1. Are $\frac{4}{5}$ and $\frac{1}{2}$ equivalent? Explain.

_____

_____

_____

2. Describe another way you could tell whether $\frac{4}{5}$ and $\frac{1}{2}$ are equivalent.

_____

_____

_____

Share and Show

1. Find a common denominator for $\frac{1}{3}$ and $\frac{1}{12}$ by dividing each whole into the same number of equal parts. Use the models to help.

common denominator: _____

$\frac{1}{3}$ $\frac{1}{12}$

© Houghton Mifflin Harcourt Publishing Company

Name _____

Write the pair of fractions as a pair of fractions with a common denominator.

2. $\frac{1}{2}$ and $\frac{1}{4}$

$\frac{2}{4}$ $\frac{2}{8}$

3. $\frac{3}{4}$ and $\frac{5}{8}$

$\frac{6}{8}$ $\frac{10}{16}$

4. $\frac{1}{3}$ and $\frac{1}{4}$

$\frac{2}{6}$ $\frac{2}{8}$

5. $\frac{4}{12}$ and $\frac{5}{8}$

$\frac{8}{24}$

Math Talk

MATHEMATICAL PRACTICES ⑥

Explain how using a model or listing multiples helps you find a common denominator.

On Your Own

Write the pair of fractions as a pair of fractions with a common denominator.

6. $\frac{1}{4}$ and $\frac{5}{6}$

$\frac{2}{8}$

7. $\frac{3}{5}$ and $\frac{4}{10}$

$\frac{6}{10}$

Tell whether the fractions are equivalent. Write = or ≠.

8. $\frac{3}{4}$ ⊘ $\frac{1}{2}$

9. $\frac{3}{4}$ = $\frac{6}{8}$

10. $\frac{1}{2}$ = $\frac{4}{8}$

11. $\frac{6}{8}$ ≠ $\frac{4}{8}$

12. **GO DEEPER** Jerry has two same-size circles divided into the same number of equal parts. One circle has $\frac{3}{4}$ of the parts shaded, and the other has $\frac{2}{3}$ of the parts shaded. His sister says the least number of pieces each circle could be divided into is 7. Is his sister correct? Explain.

_____

_____

© Houghton Mifflin Harcourt Publishing Company

Problem Solving • Applications (Real World)

13. **GO DEEPER** Carrie has a red streamer that is $\frac{3}{4}$ yard long and a blue streamer that is $\frac{5}{6}$ yard long. She says the streamers are the same length. Does this make sense? Explain.

_____

_____

_____

14. **THINK SMARTER** Leah has two same-size rectangles divided into the same number of equal parts. One rectangle has $\frac{1}{3}$ of the parts shaded, and the other has $\frac{2}{5}$ of the parts shaded. What is the least number of parts into which both rectangles could be divided?

_____

15. **MATHEMATICAL PRACTICE 6** Julian says a common denominator for $\frac{3}{4}$ and $\frac{2}{5}$ is 9. What is Julian's error? **Explain**.

WRITE *Math*
Show Your Work

_____

_____

_____

Personal Math Trainer

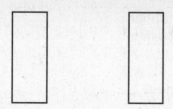

16. **THINK SMARTER +** Miguel has two same-size rectangles divided into the same number of equal parts. One rectangle has $\frac{3}{4}$ of the parts shaded, and the other has $\frac{5}{8}$ of the parts shaded.

Into how many parts could each rectangle be divided? Show your work by sketching the rectangles.

348

© Houghton Mifflin Harcourt Publishing Company

Common Denominators

Common
Core   **COMMON CORE STANDARD—4.NF.A.1**
Extend understanding of fraction equivalence and ordering.

Write the pair of fractions as a pair of fractions with a common denominator.

1. $\frac{2}{3}$ and $\frac{3}{4}$

   Think: Find a common multiple.

   3: 3, 6, 9, ⑫ 15

   4: 4, 8, ⑫ 16, 20

   $\frac{8}{12}$, $\frac{9}{12}$

2. $\frac{1}{4}$ and $\frac{2}{3}$

   $\frac{2}{8}$ $\frac{4}{6}$

3. $\frac{3}{10}$ and $\frac{1}{2}$

   $\frac{6}{20}$ $\frac{2}{4}$

4. $\frac{3}{5}$ and $\frac{3}{4}$

   $\frac{6}{10}$ $\frac{6}{8}$

5. $\frac{2}{4}$ and $\frac{7}{8}$

   $\frac{3}{6}$ $\frac{14}{16}$ $\frac{10}{10}$

6. $\frac{2}{3}$ and $\frac{5}{12}$

   $\frac{4}{6}$ $\frac{10}{24}$

7. $\frac{1}{4}$ and $\frac{1}{6}$

   $\frac{2}{8}$ $\frac{2}{12}$

Tell whether the fractions are equivalent. Write = or ≠.

8. $\frac{1}{2}$ ⬦ $\frac{2}{5}$

9. $\frac{1}{2}$ ◯ $\frac{3}{6}$

10. $\frac{3}{4}$ ◯ $\frac{5}{6}$

11. $\frac{6}{10}$ ◯ $\frac{3}{5}$

Problem Solving *(Real World)*

12. Adam drew two same size rectangles and divided them into the same number of equal parts. He shaded $\frac{1}{3}$ of one rectangle and $\frac{1}{4}$ of the other rectangle. What is the least number of parts into which both rectangles could be divided?

13. Mera painted equal sections of her bedroom wall to make a pattern. She painted $\frac{2}{5}$ of the wall white and $\frac{1}{2}$ of the wall lavender. Write an equivalent fraction for each fraction using a common denominator.

_____

14. **WRITE** ▸*Math* How are a common denominator and a common multiple alike and different?

_____

_____

© Houghton Mifflin Harcourt Publishing Company

Lesson Check (4.NF.A.1)

1. Write a common denominator for $\frac{1}{4}$ and $\frac{5}{6}$.

Answer

$$\frac{6}{24} \qquad \frac{20}{24}$$

2. Two fractions have a common denominator of 8. What could the two fractions be?

$$\frac{42}{88} \quad \frac{1}{2} \quad \frac{1}{4}$$

Spiral Review (4.NBT.A.2, 4.NBT.B.5, 4.NBT.B.6, 4.NF.A.1)

3. What number is 100,000 more than seven hundred two thousand, eighty-three?

4. Aiden baked 8 dozen muffins. How many total muffins did he bake?

5. On a bulletin board, the principal, Ms. Gomez, put 115 photos of the fourth-grade students in her school. She put the photos in 5 equal rows. How many photos did she put in each row?

6. Judy uses 12 tiles to make a mosaic. Eight of the tiles are blue. What fraction, in simplest form, represents the tiles that are blue?

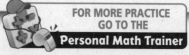

FOR MORE PRACTICE
GO TO THE
Personal Math Trainer

© Houghton Mifflin Harcourt Publishing Company

Problem Solving • Find Equivalent Fractions

Essential Question How can you use the strategy *make a table* to solve problems using equivalent fractions?

Number and Operations—
Fractions—4.NF.A.1
MATHEMATICAL PRACTICES
MP1, MP4, MP6

🔑 Unlock the Problem (Real World)

Anaya is planting a flower garden. The garden will have no more than 12 equal sections. $\frac{3}{4}$ of the garden will have daisies. What other fractions could represent the part of the garden that will have daisies?

Read the Problem

What do I need to find?	**What information do I need to use?**	**How will I use the information?**
_____ that could represent the part of the garden that will have daisies	_____ of the garden will have daisies. The garden will not have more than _____ equal sections.	I can make a _____ to find _____ fractions to solve the problem.

Solve the Problem

I can make a table and draw models to find equivalent fractions.

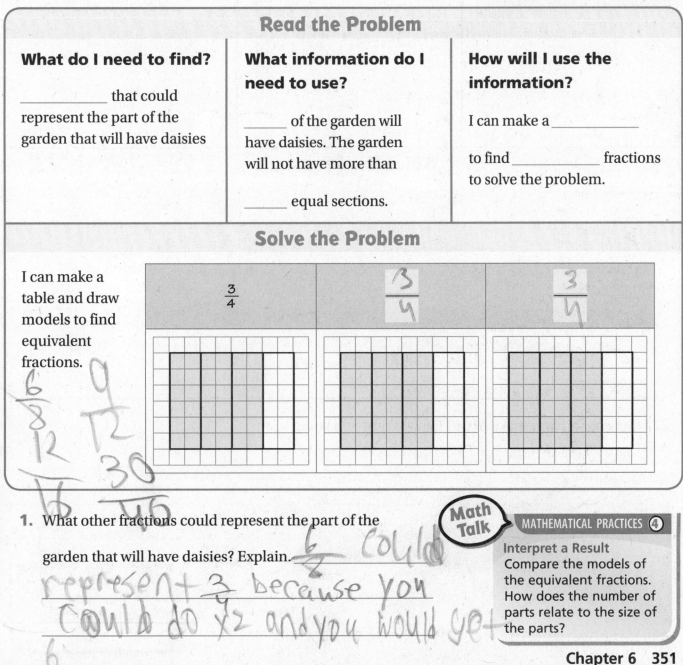

| $\frac{3}{4}$ | $\frac{3}{4}$ | $\frac{3}{4}$ |

$\frac{6}{8}$ $\frac{9}{12}$

$\frac{12}{12}$ $\frac{30}{?}$

1. What other fractions could represent the part of the garden that will have daisies? Explain. $\frac{6}{8}$ could

represent $\frac{3}{4}$ because you could do x2 and you would get

$\frac{6}{8}$

Math Talk

MATHEMATICAL PRACTICES ④

Interpret a Result
Compare the models of the equivalent fractions. How does the number of parts relate to the size of the parts?

© Houghton Mifflin Harcourt Publishing Company • Image Credits: (t) ©Ariel Skelley/Getty Images

🔒 Try Another Problem

Two friends are knitting scarves. Each scarf has 3 rectangles, and $\frac{2}{3}$ of the rectangles have stripes. If the friends are making 10 scarves, how many rectangles do they need? How many rectangles will have stripes?

Read the Problem

What do I need to find?	What information do I need to use?	How will I use the information?

Solve the Problem

2. Does your answer make sense? Explain how you know.

_____

_____

_____

Math Talk

MATHEMATICAL PRACTICES ①

Analyze What other strategy could you have used and why?

© Houghton Mifflin Harcourt Publishing Company

Name _____

Share and Show

Unlock the Problem

√ Use the Problem Solving Mathboard.
√ Underline important facts.
√ Choose a strategy you know.

1. Keisha is helping plan a race route for a 10-kilometer charity run. The committee wants to set up the following things along the course.

> **Viewing areas:** At the end of each half of the course
>
> **Water stations:** At the end of each fifth of the course
>
> **Distance markers:** At the end of each tenth of the course

Which locations have more than one thing located there?

First, make a table to organize the information.

	Number of Locations	First Location	All the Locations
Viewing Areas	2	$\frac{1}{2}$	$\frac{1}{2}$
Water Stations	5	$\frac{1}{5}$	$\frac{1}{5}$
Distance Markers	10	$\frac{1}{10}$	$\frac{1}{10}$

Next, identify a relationship. Use a common denominator, and find equivalent fractions.

Finally, identify the locations at which more than one thing will be set up. Circle the locations.

_____

_____

2. **THINK SMARTER** What if distance markers will also be placed at the end of every fourth of the course? Will any of those markers be set up at the same location as another distance marker, a water station,

or a viewing area? Explain. _____

_____

_____

3. Fifty-six students signed up to volunteer for the race. There were 4 equal groups of students, and each group had a different task.

How many students were in each group? _____

© Houghton Mifflin Harcourt Publishing Company • Image Credits: (tr) ©Stephen Coburn/Shutterstock; (cr) ©Stephen Coburn/Shutterstock; (br) ©Stephen Coburn/Shutterstock

On Your Own

4. **THINK SMARTER** A baker cut a pie in half. He cut each half into 3 equal pieces and each piece into 2 equal slices. He sold 6 slices. What fraction of the pie did the baker sell?

$$\frac{6}{6}$$

5. **GO DEEPER** Andy cut a tuna sandwich and a chicken sandwich into a total of 15 same-size pieces. He cut the tuna sandwich into 9 more pieces than the chicken sandwich. Andy ate 8 pieces of the tuna sandwich. What fraction of the tuna sandwich did he eat?

WRITE ▸ Math
Show Your Work

6. **MATHEMATICAL PRACTICE ⑥** Luke threw balls into these buckets at a carnival. The number on the bucket gives the number of points for each throw. What is the least number of throws needed to score exactly 100 points? **Explain.**

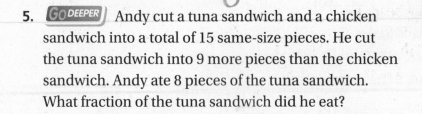

One hundre the last
digit is 0 so its even
and it 20

7. **THINK SMARTER** Victoria arranges flowers in vases at her restaurant. In each arrangement, $\frac{2}{3}$ of the flowers are yellow. What other fractions can represent the part of the flowers that are yellow? Shade the models to show your work.

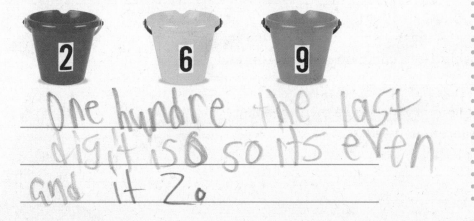

$$\frac{2}{3} \qquad \overline{12} \qquad \underline{}$$

© Houghton Mifflin Harcourt Publishing Company

Problem Solving • Find Equivalent Fractions

 COMMON CORE STANDARD—4.NF.A.1
Extend understanding of fraction equivalence and ordering.

Solve each problem.

1. Miranda is braiding her hair. Then she will attach beads to the braid. She wants $\frac{1}{3}$ of the beads to be red. If the greatest number of beads that will fit on the braid is 12, what other fractions could represent the part of the beads that are red?

   $\dfrac{2}{6}, \dfrac{3}{9}, \dfrac{4}{12}$

2. Ms. Groves has trays of paints for students in her art class. Each tray has 5 colors. One of the colors is purple. What fraction of the colors in 20 trays is purple?

   $\dfrac{4}{20}$

3. Miguel is making an obstacle course for field day. At the end of every sixth of the course, there is a tire. At the end of every third of the course, there is a cone. At the end of every half of the course, there is a hurdle. At which locations of the course will people need to go through more than one obstacle?

   hurdle

4. **WRITE** ▸*Math* Draw and compare models of $\frac{3}{4}$ of a pizza pie and $\frac{6}{8}$ of a same-size pie.

   _____

   _____

   _____

© Houghton Mifflin Harcourt Publishing Company

Lesson Check (4.NF.A.1)

1. A used bookstore will trade 2 of its books for 3 of yours. If Val brings in 18 books to trade, how many books can she get from the store?

2. Every $\frac{1}{2}$ hour Naomi stretches her neck; every $\frac{1}{3}$ hour she stretches her legs; and every $\frac{1}{6}$ hour she stretches her arms. Which parts of her body will Naomi stretch when $\frac{2}{3}$ of an hour has passed?

Spiral Review (4.OA.B.4, 4.NBT.B.4, 4.NBT.B.6, 4.NF.A.1)

3. At the beginning of the year, the Wong family car had been driven 14,539 miles. At the end of the year, their car had been driven 21,844 miles. How many miles did the Wong family drive their car during that year?

4. Widget Company made 3,600 widgets in 4 hours. They made the same number of widgets each hour. How many widgets did the company make in one hour?

5. Tyler is thinking of a number that is divisible by 2 and by 3. Write another number by which Tyler's number must also be divisible.

6. Jessica drew a circle divided into 8 equal parts. She shaded 6 of the parts. What fraction is equivalent to the part of the circle that is shaded?

 © Houghton Mifflin Harcourt Publishing Company

FOR MORE PRACTICE
GO TO THE
Personal Math Trainer

Compare Fractions Using Benchmarks

Essential Question How can you use benchmarks to compare fractions?

Common Core Number and Operations—
Fractions—4.NF.A.2

MATHEMATICAL PRACTICES
MP2, MP6, MP7

🔑 Unlock the Problem Real World

David made a popcorn snack. He mixed $\frac{5}{8}$ gallon of popcorn with $\frac{1}{2}$ gallon of dried apple rings. Did he use more dried apple rings or more popcorn?

🔑 Activity Compare $\frac{5}{8}$ and $\frac{1}{2}$.

Materials ■ fraction strips

Use fraction strips to compare $\frac{5}{8}$ and $\frac{1}{2}$. Record on the model below.

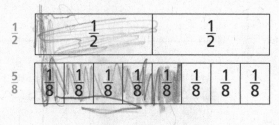

$\frac{5}{8} \bigcirc \frac{1}{2}$

So, David used more _____.

Math Talk

MATHEMATICAL PRACTICES ⑦

Look for Structure How are the number of eighth-size parts in $\frac{5}{8}$ related to the number of eighth-size parts you need to make $\frac{1}{2}$?

1. Write five fractions equivalent to $\frac{1}{2}$. What is the relationship between the numerator and the denominator of fractions equivalent to $\frac{1}{2}$?

for 2 12's myltiples reach up
to y that 1

$\frac{3}{4}$ $\frac{3}{6}$ $\frac{4}{8}$ $\frac{5}{10}$

$\frac{6}{12}$

2. How many eighths are equivalent to $\frac{1}{2}$?

$\frac{4}{8}$

3. How can you compare $\frac{5}{8}$ and $\frac{1}{2}$ without using a model?

$\frac{5}{8}$ would be bigger than $\frac{1}{2}$

© Houghton Mifflin Harcourt Publishing Company • Image Credits: (t) ©Image Source/Getty Images

Benchmarks A **benchmark** is a known size or amount that helps you understand a different size or amount. You can use $\frac{1}{2}$ as a benchmark to help you compare fractions.

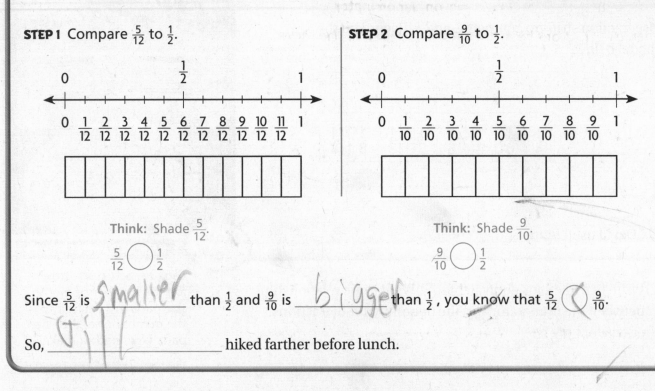

Example Use benchmarks to compare fractions.

A family hiked the same mountain trail. Evie and her father hiked $\frac{5}{12}$ of the trail before they stopped for lunch. Jill and her mother hiked $\frac{9}{10}$ of the trail before they stopped for lunch. Who hiked farther before lunch?

Compare $\frac{5}{12}$ and $\frac{9}{10}$ to the benchmark $\frac{1}{2}$.

STEP 1 Compare $\frac{5}{12}$ to $\frac{1}{2}$.

0     $\frac{1}{2}$ 1

0   $\frac{1}{12}$ $\frac{2}{12}$ $\frac{3}{12}$ $\frac{4}{12}$ $\frac{5}{12}$ $\frac{6}{12}$ $\frac{7}{12}$ $\frac{8}{12}$ $\frac{9}{12}$ $\frac{10}{12}$ $\frac{11}{12}$ 1

Think: Shade $\frac{5}{12}$.

$\frac{5}{12}$ ◯ $\frac{1}{2}$

STEP 2 Compare $\frac{9}{10}$ to $\frac{1}{2}$.

0     $\frac{1}{2}$ 1

0   $\frac{1}{10}$ $\frac{2}{10}$ $\frac{3}{10}$ $\frac{4}{10}$ $\frac{5}{10}$ $\frac{6}{10}$ $\frac{7}{10}$ $\frac{8}{10}$ $\frac{9}{10}$ 1

Think: Shade $\frac{9}{10}$.

$\frac{9}{10}$ ◯ $\frac{1}{2}$

Since $\frac{5}{12}$ is _smaller_ than $\frac{1}{2}$ and $\frac{9}{10}$ is _bigger_ than $\frac{1}{2}$, you know that $\frac{5}{12}$ ⬤ $\frac{9}{10}$.

So, _Jill_ hiked farther before lunch.

4. Explain how you can tell $\frac{5}{12}$ is less than $\frac{1}{2}$ without using a model.

   5/12 is smaller than 1/2 becaus six twelves is half of twelve.

5. Explain how you can tell $\frac{7}{10}$ is greater than $\frac{1}{2}$ without using a model.

   seven tents is greater than one half because five tenths is half of 10.

360

© Houghton Mifflin Harcourt Publishing Company • Image Credits: (t) ©Steve Mason/Getty Images

Name _____

Compare Fractions

Essential Question How can you compare fractions?

Common Core **Number and Operations— Fractions—4.NF.A.2**
MATHEMATICAL PRACTICES
MP2, MP3, MP6

🔑 Unlock the Problem Real World

Every year, Avery's school has a fair. This year, $\frac{3}{8}$ of the booths had face painting and $\frac{1}{4}$ of the booths had sand art. Were there more booths with face painting or sand art?

Compare $\frac{3}{8}$ and $\frac{1}{4}$.

🔑 One Way Use a common denominator.

When two fractions have the same denominator, they have equal-size parts. You can compare the number of parts.

THINK

Think: 8 is a multiple of both 4 and 8.
Use 8 as a common denominator.

$$\frac{1}{4} = \frac{1 \times 2}{4 \times 2} = \frac{2}{8}$$

$\frac{3}{8}$ already has 8 as a denominator.

MODEL AND RECORD

Shade the model. Then compare.

$\frac{3}{8}$ ⊘ $\frac{2}{8}$

🔑 Another Way Use a common numerator.

When two fractions have the same numerator, they represent the same number of parts. You can compare the size of the parts.

THINK

Think: 3 is a multiple of both 3 and 1.
Use 3 as a common numerator.

$\frac{3}{8}$ already has 3 as a numerator.

$$\frac{1}{4} = \frac{1 \times 3}{4 \times 3} = \frac{3}{12}$$

MODEL AND RECORD

Shade the model. Then compare.

$\frac{3}{8}$ ◯ $\frac{3}{12}$

Since $\frac{3}{8}$ ◯ $\frac{1}{4}$, there were more booths with _____.

Math Talk

MATHEMATICAL PRACTICES ②

Reason Abstractly Why can you not use $\frac{1}{2}$ as a benchmark to compare $\frac{3}{8}$ and $\frac{1}{4}$?

© Houghton Mifflin Harcourt Publishing Company • Image Credits: (t) © Doug Menuez/Getty Images

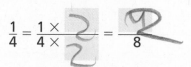

Try This! Compare the fractions. Explain your reasoning.

Ⓐ $\frac{3}{4}$ ⃠ $\frac{1}{3}$

$\frac{9}{12}$ $\frac{4}{12}$

If you have 4's multiples and 3's they meet at 12 and the 4x1=4 and 3x3 is 9.

Ⓑ $\frac{3}{5}$ ⃠ $\frac{3}{8}$ 15

$\frac{24}{40}$ $\frac{15}{40}$

Five and eight meet at forty and five times three equls fifteen and eight times three is twenty four.

Ⓒ $\frac{3}{4}$ ⃠ $\frac{7}{8}$ 28

$\frac{24}{32}$ $\frac{28}{32}$

Four and eight meet at thirtytwo and four times seven is twenty eight and eight times three is twenty four.

Ⓓ $\frac{4}{5}$ ⃠ $\frac{2}{3}$ 10

$\frac{12}{15}$ $\frac{10}{15}$

Five and three's multip meet at fiftheen and five time two is ten and four times three is twelve.

1. Which would you use to compare $\frac{11}{12}$ and $\frac{5}{6}$, a common numerator or a common denominator? Explain.

Six times two is twelve but five times two is ten not eleven.

2. Can you use simplest form to compare $\frac{8}{10}$ and $\frac{3}{5}$? Explain.

Divide two by eight and ten It would be four fifths because you can't divide five. Three fifths both you can't divide

$\frac{8}{10} = \frac{4}{5}$

$\frac{3}{5}$

© Houghton Mifflin Harcourt Publishing Company

Share and Show MATH BOARD

1. Compare $\frac{2}{5}$ and $\frac{1}{10}$.

   Think: Use ___11___ as a common denominator.

   $\frac{2}{5} = \frac{2 \times 2}{5 \times 2} = \frac{4}{10}$

   $\frac{1}{10}$

   Think: 4 tenth-size parts ⊘ 1 tenth-size part.

   $\frac{2}{5}$ ⊘ $\frac{1}{10}$

2. Compare $\frac{6}{10}$ and $\frac{3}{4}$.

   Think: Use _____ as a common numerator.

   $\frac{6}{10}$

   $\frac{3}{4} = \frac{3 \times 2}{4 \times 2} = \frac{6}{8}$

   Think: A tenth-size part ◯ an eighth-size part.

   $\frac{6}{10}$ ◯ $\frac{3}{4}$

Compare. Write <, >, or =.

3. $\frac{7}{8}$ ◯ $\frac{2}{8}$

4. $\frac{5}{12}$ ◯ $\frac{3}{6}$

5. $\frac{4}{10}$ ◯ $\frac{4}{6}$

6. $\frac{6}{12}$ ◯ $\frac{2}{4}$

Math Talk MATHEMATICAL PRACTICES ②

Use Reasoning How can using a common numerator or a common denominator help you compare fractions?

On Your Own

Compare. Write <, >, or =.

7. $\frac{1}{3}$ ◯ $\frac{1}{4}$

8. $\frac{4}{5}$ ◯ $\frac{8}{10}$

9. $\frac{3}{4}$ ◯ $\frac{2}{6}$

10. $\frac{1}{2}$ ◯ $\frac{5}{8}$

MATHEMATICAL PRACTICE ② Reason Quantitatively **Algebra** Find a number that makes the statement true.

11. $\frac{1}{2} > \frac{1}{3}$

12. $\frac{3}{10} < \frac{4}{5}$

13. $\frac{5}{12} < \frac{2}{3}$

14. $\frac{2}{3} > \frac{4}{3}$

15. **GO DEEPER** Students cut a pepperoni pizza into 12 equal slices and ate 5 slices. They cut a veggie pizza into 6 equal slices and ate 4 slices. Use fractions to compare the amounts of each pizza that were eaten.

$\frac{5}{12} < \frac{4}{6}$

© Houghton Mifflin Harcourt Publishing Company

🔑 Unlock the Problem (Real World)

16. **THINK SMARTER** Jerry is making a strawberry smoothie. Which measure is greatest, the amount of milk, cottage cheese, or strawberries?

Strawberry Smoothie

3 ice cubes

$\frac{3}{4}$ cup milk

$\frac{2}{6}$ cup cottage cheese

$\frac{8}{12}$ cup strawberries

$\frac{1}{4}$ teaspoon vanilla

$\frac{1}{8}$ teaspoon sugar

a. What do you need to find?

The cottage cheese, strawberries, and milk

b. How will you find the answer?

See what fractions of cottage cheese, milk, strawber

c. Show your work.

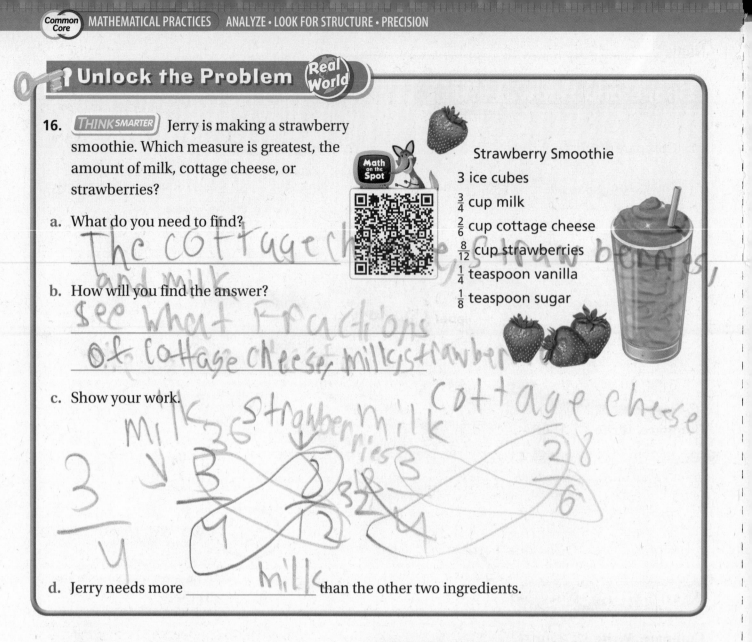

d. Jerry needs more ___milk___ than the other two ingredients.

---

17. **GO DEEPER** Angie, Blake, Carlos, and Daisy went running. Angie ran $\frac{1}{3}$ mile, Blake ran $\frac{3}{5}$ mile, Carlos ran $\frac{7}{10}$ mile, and Daisy ran $\frac{1}{2}$ mile. Which runner ran the shortest distance? Who ran the greatest distance?

18. **THINK SMARTER** Elaine bought $\frac{5}{8}$ pound of potato salad and $\frac{4}{6}$ pound of macaroni salad for a picnic. Use the numbers to compare the amounts of potato salad and macaroni salad Elaine bought.

☐ / ☐ < ☐ / ☐

| 4 |
| 5 |
| 6 |
| 8 |

© Houghton Mifflin Harcourt Publishing Company

Compare and Order Fractions

Essential Question How can you order fractions?

Common Core Number and Operations—
Fractions—4.N .2
MATHEMATICAL PRACTICES
MP2, MP4, MP6

🔑 Unlock the Problem *Real World*

Jody has equal-size bins for the recycling center. She filled $\frac{3}{5}$ of a bin with plastics, $\frac{1}{12}$ of a bin with paper, and $\frac{9}{10}$ of a bin with glass. Which bin is the most full?

- Underline what you need to find.
- Circle the fractions you will compare.

🔒 Example 1 Locate and label $\frac{3}{5}$, $\frac{1}{12}$, and $\frac{9}{10}$ on the number line.

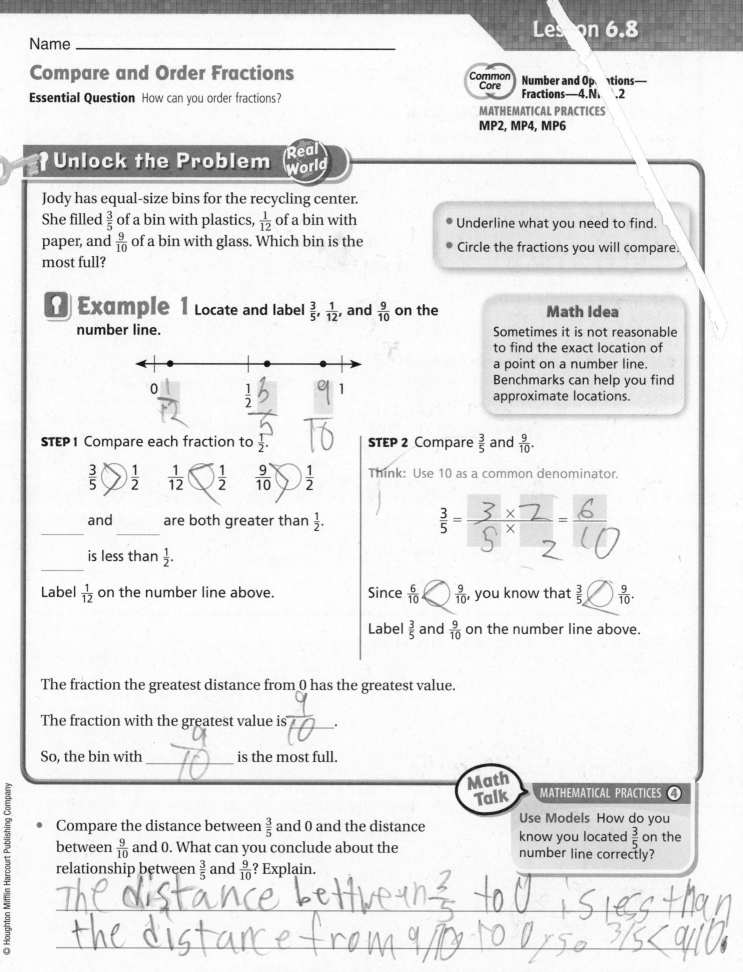

Math Idea
Sometimes it is not reasonable to find the exact location of a point on a number line. Benchmarks can help you find approximate locations.

STEP 1 Compare each fraction to $\frac{1}{2}$.

$$\frac{3}{5} \bigcirc \frac{1}{2} \qquad \frac{1}{12} \bigcirc \frac{1}{2} \qquad \frac{9}{10} \bigcirc \frac{1}{2}$$

_____ and _____ are both greater than $\frac{1}{2}$.

_____ is less than $\frac{1}{2}$.

Label $\frac{1}{12}$ on the number line above.

STEP 2 Compare $\frac{3}{5}$ and $\frac{9}{10}$.

Think: Use 10 as a common denominator.

$$\frac{3}{5} = \frac{3 \times 2}{5 \times 2} = \frac{6}{10}$$

Since $\frac{6}{10} \bigcirc \frac{9}{10}$, you know that $\frac{3}{5} \bigcirc \frac{9}{10}$.

Label $\frac{3}{5}$ and $\frac{9}{10}$ on the number line above.

The fraction the greatest distance from 0 has the greatest value.

The fraction with the greatest value is $\frac{9}{10}$.

So, the bin with $\frac{9}{10}$ is the most full.

- Compare the distance between $\frac{3}{5}$ and 0 and the distance between $\frac{9}{10}$ and 0. What can you conclude about the relationship between $\frac{3}{5}$ and $\frac{9}{10}$? Explain.

The distance between 3/5 to 0 is less than the distance from 9/10 to 0 so 3/5 < 9/10.

Math Talk MATHEMATICAL PRACTICES ④
Use Models How do you know you located $\frac{3}{5}$ on the number line correctly?

© Houghton Mifflin Harcourt Publishing Company

Example 2 Write $\frac{7}{10}$, $\frac{1}{3}$, $\frac{7}{12}$, and $\frac{8}{10}$ in order from least to greatest.

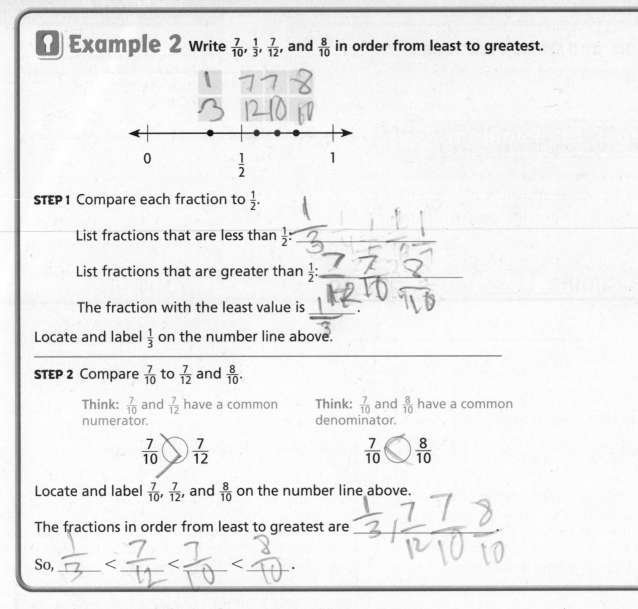

STEP 1 Compare each fraction to $\frac{1}{2}$.

List fractions that are less than $\frac{1}{2}$: _____

List fractions that are greater than $\frac{1}{2}$: _____

The fraction with the least value is _____.

Locate and label $\frac{1}{3}$ on the number line above.

STEP 2 Compare $\frac{7}{10}$ to $\frac{7}{12}$ and $\frac{8}{10}$.

Think: $\frac{7}{10}$ and $\frac{7}{12}$ have a common numerator.

$$\frac{7}{10} \bigcirc \frac{7}{12}$$

Think: $\frac{7}{10}$ and $\frac{8}{10}$ have a common denominator.

$$\frac{7}{10} \bigcirc \frac{8}{10}$$

Locate and label $\frac{7}{10}$, $\frac{7}{12}$, and $\frac{8}{10}$ on the number line above.

The fractions in order from least to greatest are _____.

So, _____ < _____ < _____ < _____.

Try This! Write $\frac{3}{4}$, $\frac{3}{6}$, $\frac{1}{3}$, and $\frac{2}{12}$ in order from least to greatest.

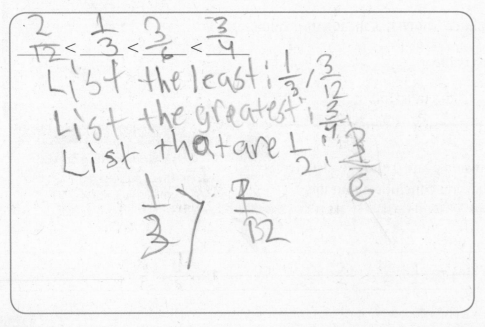

© Houghton Mifflin Harcourt Publishing Company

Name _____

1. Locate and label points on the number line to help you write
$\frac{3}{10}, \frac{11}{12},$ and $\frac{5}{8}$ in order from least to greatest.

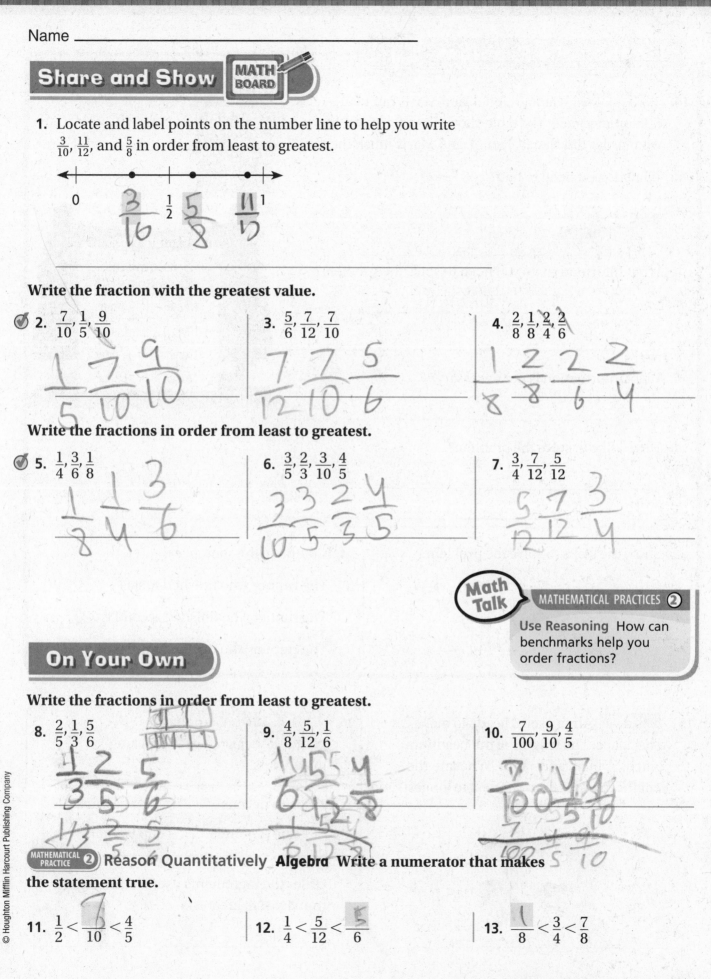

0    $\frac{3}{16}$ $\frac{1}{2}$ $\frac{5}{8}$ $\frac{11}{5}$ 1

Write the fraction with the greatest value.

2. $\frac{7}{10}, \frac{1}{5}, \frac{9}{10}$

$\frac{1}{5}$ $\frac{7}{10}$ $\frac{9}{10}$

3. $\frac{5}{6}, \frac{7}{12}, \frac{7}{10}$

$\frac{7}{12}$ $\frac{7}{10}$ $\frac{5}{6}$

4. $\frac{2}{8}, \frac{1}{8}, \frac{2}{4}, \frac{2}{6}$

$\frac{1}{8}$ $\frac{2}{8}$ $\frac{2}{6}$ $\frac{2}{4}$

Write the fractions in order from least to greatest.

5. $\frac{1}{4}, \frac{3}{6}, \frac{1}{8}$

$\frac{1}{8}$ $\frac{1}{4}$ $\frac{3}{6}$

6. $\frac{3}{5}, \frac{2}{3}, \frac{3}{10}, \frac{4}{5}$

$\frac{3}{10}$ $\frac{3}{5}$ $\frac{2}{3}$ $\frac{4}{5}$

7. $\frac{3}{4}, \frac{7}{12}, \frac{5}{12}$

$\frac{5}{12}$ $\frac{7}{12}$ $\frac{3}{4}$

Math Talk MATHEMATICAL PRACTICES ②

Use Reasoning How can benchmarks help you order fractions?

Write the fractions in order from least to greatest.

8. $\frac{2}{5}, \frac{1}{3}, \frac{5}{6}$

$\frac{1}{3}$ $\frac{2}{5}$ $\frac{5}{6}$

9. $\frac{4}{8}, \frac{5}{12}, \frac{1}{6}$

$\frac{1}{6}$ $\frac{5}{12}$ $\frac{4}{8}$

10. $\frac{7}{100}, \frac{9}{10}, \frac{4}{5}$

$\frac{7}{100}$ $\frac{4}{5}$ $\frac{9}{10}$

MATHEMATICAL PRACTICE ② **Reason Quantitatively** **Algebra** Write a numerator that makes the statement true.

11. $\frac{1}{2} < \frac{\boxed{}}{10} < \frac{4}{5}$

12. $\frac{1}{4} < \frac{5}{12} < \frac{\boxed{}}{6}$

13. $\frac{\boxed{}}{8} < \frac{3}{4} < \frac{7}{8}$

© Houghton Mifflin Harcourt Publishing Company

Unlock the Problem Real World

14. **THINK SMARTER** Nancy, Lionel, and Mavis ran in a 5-kilometer race. The table shows their finish times. In what order did Nancy, Lionel, and Mavis finish the race?

a. What do you need to find?

the racing from each person

b. What information do you need to solve the problem?

the results from each person

c. What information is not necessary?

they ran 5 kalomimforso

d. How will you solve the problem?

List all the fractions

5-Kilometer Race Results	
Name	**Time**
Nancy	$\frac{2}{3}$ hour
Lionel	$\frac{7}{12}$ hour
Mavis	$\frac{3}{4}$ hour

Finish line

e. Show the steps to solve the problem.

Mavis $\frac{3}{4}$ Nancy $\frac{2}{3}$ Lionel $\frac{7}{12}$

f. Complete the sentences.

The runner who finished first is Mavis.

The runner who finished second is Nancy

The runner who finished third is Lionel

15. **GO DEEPER** Alma used 3 beads to make a necklace. The lengths of the beads are $\frac{5}{6}$ inch, $\frac{5}{12}$ inch, and $\frac{1}{3}$ inch. What are the lengths in order from shortest to longest?

$\frac{1}{3}$ $\frac{5}{12}$ $\frac{5}{6}$

16. **THINK SMARTER** Victor has his grandmother's recipe for making mixed nuts.

$\frac{3}{4}$ cup pecans	$\frac{2}{12}$ cup peanuts
$\frac{1}{2}$ cup almonds	$\frac{7}{8}$ cup walnuts

Order the ingredients used in the recipe from least to greatest.

$\frac{2}{12}$ $\frac{1}{2}$ $\frac{3}{4}$ $\frac{7}{8}$

© Houghton Mifflin Harcourt Publishing Company